DOUGLAS DC-9 and McDONNELL DOUGLAS MD-80

P. R. SMITH

First published in the United Kingdom in 1987 by
Jane's Publishing Company Limited
238 City Road, London EC1V 2PU
in conjunction with DPR Marketing and Sales
37 Heath Road, Twickenham, Middlesex TW1 4AW

ISBN 0 7106 0427 0

 Distributed in the Philippines and the USA and its dependencies by
Jane's Publishing Inc
115 5th Avenue
New York NY 10003

Printed in the United Kingdom by Netherwood Dalton & Co Ltd

Cover illustrations

Front: **New York Air** (NY)
New York Air was established in 1980 to operate in direct competition with Eastern Airlines' 'Air Shuttle' service. The airline provides scheduled passenger services in the Northeast, Midwest, Southeast and Florida. With a main operating base at New York, New York Air operates a fleet of DC-9, MD-80 and Boeing 737 aircraft. An MD-80 is seen here arriving after a routine flight. *(P Hornfeck)*

Rear: **Scandinavian Airline System - SAS** (SK)
SAS is the national carrier of Denmark, Norway and Sweden. The airline has one of the world's largest international networks, linking more than 80 points in Scandinavia, the rest of Europe, Africa, the Middle East, Southern and Eastern Asia, North and South America. The company has a large fleet that encompasses the DC-8, DC-9, MD-80 series, DC-10 and Fokker F-27 aircraft. A SAS DC-9 can be seen here. *(Udo and Birgit Schaefer Collection)*

Right: **BWIA International** (BW)
BWIA International, the government-controlled national airline of Trinidad and Tobago, was established in 1940 as British West Indian Airways, and initially operated between Trinidad, Tobago and Barbados. The airline provides scheduled passenger services domestically, and to foreign points in the Caribbean, North and South America, as well as to Europe. The carrier has a fleet of L-1011 Tristar Srs 500, MD-83 and BAe 748 aircraft. Although no longer operated, 9Y-TFF, a DC-9-51, can be seen here in flight.
(McDonnell Douglas)

Introduction

In 1962 Douglas released design study data on a twin-engined jet aircraft. At the time the project was known as the Douglas Model 2086. Preliminary design work began that year, with fabrication beginning on 26 July 1963. The assembly of the first airframe began some eight months later. An inaugural flight was made on 25 February 1965, and five examples were flying by the end of the year. These aircraft were of the DC-9-10 Model 11, and were able to seat up to a maximum of 90 passengers. The initial version was powered by Pratt and Whitney JT8D-5 turbofan engines and went into scheduled airline service with Delta Airlines on 8 December 1965.

No fewer than 13 different models have since flown. A Series 10 Model 15 came next. This was generally similar to the original, but had upgraded engines. The Series 20 was developed for operation in hot/high conditions. Combining the long wing span of the DC-9-30, but with the short fuselage of the DC-9-10, the aircraft carried 90 passengers. The first example was delivered to SAS in 1968. The DC-9-30 was a developed version, with increased engine capacity, and with a passenger load of 115. Eastern Airlines took delivery of the first of this type in 1967. The Series 40 once again had upgraded engines and a capacity of 125 passengers. SAS took delivery of their first machine in 1968. The Series 50 had a 'new look' interior, which featured enclosed overhead racks as well as many other improvements. High density seating for 139 passengers was possible. Each type was available in passenger, cargo (DC-9F), convertible (DC-9CF) or passenger/cargo (DC-9RC) configuration. Three military versions were made available: the C-9A Nightingale, an Aeromedical airlift transport; the C-9B Skytrain II, a fleet logistic support aircraft; and the VC-9C, a specially configured type for the Special Air Mission Wing of the United States Air Force.

The MD-80 series was announced in the late 1970s. Model MD-81 went into service with Swissair in 1980. It has a 4.70 metre (15 ft 4 in) stretch over the DC-9-50. The MD-82 has upgraded engines suitable for use from hot and high airfields. Model MD-83 has upgraded engines and an increase in range. The MD-87, a short-fuselage MD-80 version, is due to go into service in late 1987 with Austrian Airlines and Finnair.

I would like to extend my sincere thanks to everyone who has contributed in one way or another in the preparation of this book.

Finally, I would like to dedicate this book to Joan Penny, a very dear and special friend.

TABLE OF COMPARISONS

	DC-9-10	DC-9-20	DC-9-30
Max. accommodation	90	90	115
Wing span	27.25 m (89 ft 5 in)	28.47 m (93 ft 5 in)	28.47 m (93 ft 5 in)
Length	31.82 m (108 ft 4.75 in)	31.82 m (108 ft 4.75 in)	31.82 m (108 ft 4.75 in)
Height	8.38 m (27 ft 6 in)	8.38 m (27 ft 6 in)	8.38 m (27 ft 6 in)
Max. t/o weight	41 140 kg (90 700 lb)	44 450 kg (98 000 lb)	44 450 kg (98 000 lb)
Max. cruis. speed	903 km/h (561 mph)	903 km/h (561 mph)	909 km/h (565 mph)
Maximum range	1601 km (995 miles)	2970 km (1843 miles)	2775 km (1725 miles)
	DC-9-40	**DC-9-50**	**MD-81**
Max. accommodation	125	139	172
Wing span	28.47 m (93 ft 5 in)	28.47 m (93 ft 5 in)	32.87 m (107 ft 10 in)
Length	38.28 m (125 ft 7.25 in)	40.72 m (133 ft 7.25 in)	45.06 m (147 ft 10 in)
Height	8.53 m (28 ft 0 in)	8.53 m (28 ft 0 in)	9.04 m (29 ft 8 in)
Max. t/o weight	51 710 kg (114 000 lb)	54 885 kg (121 000 lb)	63 503 kg (140 000 lb)
Max. cruis. speed	903 km/h (561 mph)	898 km/h (558 mph)	925 km/h (575 mph)
Maximum range	2710 km (1685 miles)	4066 km (2527 miles)	2896 km (1800 miles)
	MD-82	**MD-83**	
Max. accommodation	172	172	
Wing span	32.87 m (107 ft 10 in)	32.87 m (107 ft 10 in)	
Length	45.06 m (147 ft 10 in)	45.06 m (147 ft 10 in)	
Height	9.04 m (29 ft 8 in)	9.04 m (29 ft 8 in)	
Max. t/o weight	67 812 kg (149 500 lb)	72 575 kg (160 000 lb)	
Max. cruis. speed	925 km/h (575 mph)	925 km/h (575 mph)	
Maximum range	3798 km (2360 miles)	4395 km (2731 miles)	

Opposite: **East African Airways Corporation** (EC)
EAAC was established in 1945, with operations commencing on 1 January 1946. The company was the joint national airline of Kenya, Tanzania and Uganda. An extensive network of scheduled services were operated within the three partner states of the East African Community. Other international destinations included Aden, Addis Ababa, Blantyre, Bombay, London (Heathrow), Rome and Zürich. Passenger and cargo charters were operated by Simbair Ltd, a wholly-owned subsidiary. In January 1977, due to political unrest, EAAC ceased all operations. At that time, a fleet of Super VC-10, DC-9-30, F-27 and DC-3 aircraft was utilised. *(McDonnell Douglas)*

EAST AFRICAN
5H-MOI

adria aviopromet
DC-9

Opposite: **Adria Aviopromet** (JP)
Until recently known as Inex Adria, Aviopromet Airways is Yugoslavia's second largest carrier. The airline was formed in 1960 and began operations a year later, using DC-6B aircraft. In 1969 the airline became part of the Interexport organisation and changed its name to Inex Adria Airways. With this change, the company acquired its first DC-9-32 aircraft. Today, scheduled passenger services are operated to various parts of Yugoslavia. Group charter operations link the country with points throughout Europe, North Africa and the Middle East. Adria operates a fleet of DC-9, MD-80 and de Havilland Canada Dash 7 aircraft. A DC-9-32, YU-AHW, is seen here just after push-back at London (Gatwick) Airport. *(J Page)*

Above: **Aermediterranea** (BQ)
Aermediterranea was formed in March 1981 as a subsidiary of Alitalia, and to take over many routes of the then recently-defunct airline, Itavia. An inaugural service was made over a Rome-Lameriza Terme route in July of that year. Ownership of the airline was held by the Italian government through Alitalia and ATI, as well as various private interests. In mid-1985, a corporate merger between Aermediterranea and ATI was made. The airline, based at Naples, undertook various domestic scheduled and international charter activities. Aermediterranea operated a fleet of DC-9-30 aircraft. *(Udo and Birgit Schaefer Collection)*

Aero Lloyd (LL)

Aero Lloyd, a privately-owned West German charter company, was established in 1979. From its base at Frankfurt, the carrier made its inaugural flight on 1 April 1981, using Caravelle equipment. The airline operates international tour group flights within Europe, and to the Middle East and North Africa. The company operates a fleet of DC-9, MD-83 and ageing Caravelle aircraft. D-ALLD, its MD-83, is seen here at Frankfurt.
(McDonnell Douglas)

Aeroméxico (AM)
Aeroméxico, Mexico's government-controlled national flag carrier, operates extensive domestic services to over 38 points, as well as international services to the USA, Canada, Panama, Colombia, Venezuela, France and Spain. The airline was established in September 1934 as Aeronaves de México, and initially began services over a Mexico City-Acapulco route, using a single-engined Stinson aircraft. Today Aeroméxico is Mexico's second largest carrier in terms of passengers carried and operates a fleet of DC-9, DC-10 and MD-80 aircraft. A DC-9-32 is seen here.
(Udo and Birgit Schaefer Collection)

Aeropostal Airlines (LV)
Aeropostal, the government-controlled Venezuelan airline, is a well-developed company that operates both regional international and domestic scheduled flights. The carrier was established as Compagnie Generale Aeropostale in 1930. Following nationalisation in 1933, its name was changed to Linea Aeropostale Venezolana — or Aeropostal for short. In mid-1957, operations were expanded when the airline TACA de Venezuela was completely absorbed. Today, international services are operated to the Netherlands Antilles and to Port of Spain in Trinidad and Tobago. Aeropostal operates a fleet of DC-9 and DHC-6 Twin Otter aircraft. A DC-9-51 can be seen here. *(McDonnell Douglas)*

Airborne Express (GB)

Airborne Express is the airline division of Airborne Freight Corporation, one of the leading package express and freight forwarding companies in the USA. The company was formed in April 1980 when Airborne Freight acquired Midwest Air Charter. Midwest had at one time flown contract services for the company, and its aircraft were able to form the backbone of the initial Airborne Express fleet. Flights are operated throughout the USA as well as international charter services. At night, flights from throughout the system converge on the Wilmington hub, the company's base, where packages are sorted and transferred to outbound aircraft departing in the early morning. A fleet of DC-9 and YS-11 aircraft is operated. N903AX, a DC-9-32, is seen here at Miami. *(N Chalcroft)*

AIR CANADA
DC-9

Opposite: **Air Canada** (AC)
Air Canada was formed as Trans Canada Airlines by the Canadian government in 1937. The company's inaugural service was over a Vancouver-Seattle route (an airmail run which was taken over from Canadian Airways) on 1 September of that year. Today Air Canada, the government-controlled Canadian flag carrier, maintains scheduled jet services to over 32 destinations in Canada and to international points in the USA, Bermuda, the Bahamas, the Caribbean, Europe and southern Asia, ranking among the world's largest airlines. Air Canada's aviation holdings include an 85.6 per cent interest in Nordair, and a 30 per cent share in Innotech Aviation. The current fleet complement includes Boeing 727, 747, 767, L-1011 Tristar, and DC-9-32 aircraft. An early DC-9-10 is seen here in-flight. *(McDonnell Douglas)*

Below: **Air Florida** (QH)
Air Florida, a wholly-owned subsidiary of Air Florida System, was formed in September 1971 as a scheduled interstate passenger carrier. Operations began a year later and high-frequency, low-fare services linked Miami with the rest of Florida. Interstate services were also operated to New York, Washington DC, Philadelphia, Toledo and Houston. International services reached Europe, the Bahamas, Jamaica and the Turks and Caicos Islands. In 1984, due to financial difficulties, Air Florida ceased operations and filed for bankruptcy. At that time the airline was operating a fleet of DC-9, DC-10, Boeing 737 and DHC-6 Twin Otter aircraft. *(D Penny)*

Alaska Airlines (AS)
Alaska Airlines can trace its history as far back as 1932, when McGee Airways commenced an Anchorage-Bristol Bay service with a 3-seater Stinson aircraft. The present name was adopted in 1964, following a series of mergers. In 1951 the company began a service over a Fairbanks-Anchorage-Seattle line, using DC-4s. Convair 880 jet aircraft were introduced 10 years later, and in 1968 Alaska Airlines absorbed two important regional carriers, Cordova Airlines and Alaska Coastal Airlines. Today, the carrier maintains a scheduled passenger network that connects over 90 points. Services include interstate flights as well as to the Pacific, Northwest and California. A fleet of Boeing 727, 737 and MD-83 aircraft is utilised. *(N Chalcroft)*

Alisarda (IG)

Alisarda, a privately-owned Sardinian carrier, was established in March 1963. Charter flights commenced during the middle of the following year, with scheduled services beginning in 1966 connecting Olbia with Rome and Milan. Today, Alisarda is an expanding carrier, and flies scheduled jet services between 11 points in Italy, as well as to regional international cities in Switzerland and France. A fleet of DC-9 and MD-80 aircraft is utilised. *(Udo and Birgit Schaefer Collection)*

BJ
Alitalia
DC-9
I-DIBJ

Opposite: **Alitalia** (AZ)
Alitalia, the national airline of Italy, was established in September 1946 as Aerolinee Italiane Internazionali. Six months later an inaugural service was operated over a Turin-Rome-Catania route. The airline is today a leading international carrier. It maintains scheduled passenger and cargo services to over 70 points in Italy, the rest of Europe, Africa, the Middle East, Southern and Eastern Asia, Australia, North and South America. Alitalia ownership is controlled by the vast Italian state holding, IRI, and by several banks and investment groups. The airline owns two subsidiaries, both domestic Italian airlines, ATI-Aero Transporti Italiani and Aermediterranea, which have now merged. The company also owns real estate, insurance and hotel investment companies. Alitalia owns a fleet of Boeing 747, A300B4, MD-80, DC-9-32 and ATR 42 aircraft. A DC-9 is seen here on arrival at London (Heathrow). *(J Page)*

Above: **ALM Antillean Airlines** (LM)
ALM was founded in 1964 as a subsidiary of KLM Royal Dutch Airlines. Flight activities began the same year, utilising three Convair 340 aircraft. In 1969 the Dutch Antillean government acquired a majority shareholding in the carrier. Ownership is now split 96 per cent government and 4 per cent KLM. Today ALM flies scheduled passenger services in the Netherlands Antilles, and to Colombia, Dominican Republic, Guyana, Haiti, Jamaica, Puerto Rico, Surinam, Trinidad and Tobago, the USA and Venezuela. A fleet of DC-9 and MD-80 aircraft is used. A DC-9-32, PJ-SNA, is seen here on a visit to Miami. *(N Chalcroft)*

Below: **American International Airways (NI)**
American International started operations as a charter company in September 1981, having been formed as a successor to Commercial Airlines (a charter carrier certificated a year earlier which never began operations). Scheduled public charter services were inaugurated 13 months later followed by regular scheduled flights on 6 December 1982. Initially, the company had a flight hub at Atlantic City, but by early 1984 most services were handled at Philadelphia where routes from Boston, the Midwest and Florida converged. Unfortunately on 14 September of that year, AIA grounded all its fleet of DC-9 aircraft and suspended operations. The company had been plagued with financial difficulties. *(N Chalcroft)*

Opposite: **Ansett Airlines of Australia (AN)**
Ansett Airlines can trace its history back to 1936, and the formation of Ansett Airways. In 1957, the company purchased Australian National Airways and formed Ansett-ANA (a title which was retained until 1969). Today, the carrier is one of Australia's two primary domestic airlines. The company provides scheduled flights to over 20 points, with route concentration in the eastern states of New South Wales, Victoria, Queensland and Tasmania. Ansett operates a fleet of Boeing 727, 737, 767 and Fokker F-27 aircraft. Although no longer operated, the airline used to utilise Douglas DC-9-32s. One example is seen here in the carrier's old livery. *(McDonnell Douglas)*

ANSETT
AIRLINES OF AUSTRALIA
VH-CZG
DOUGLAS DC-9

Australian Airlines (TN)

The recently re-named Australian Airlines was formed in 1945 by the Australian National Airlines Act as Trans Australia Airlines. An inaugural flight was made over a Melbourne-Sydney line on 9 September 1946, utilising DC-3 aircraft. The airline today is one of Australia's two primary domestic air carriers, maintaining comprehensive scheduled flights between 32 points, also covering an international route from Hobart to Christchurch (New Zealand). The carrier is fully government-owned, and is run by an appointed Australian National Airlines Commission. Australian Airlines operates a fleet of DC-9, Boeing 727, 737, F-27 and A300 aircraft. A DC-9-30 is seen here sporting the livery of Trans Australia Airlines.
(Udo and Birgit Schaefer Collection)

Aviaco (AO)

Aviaco is Spain's second largest airline. It was established in 1948. Current ownership is shared by the Spanish government, Iberia, and private interests. The carrier flies extensive domestic scheduled and international tour charter services. Aviaco's primary areas of service include Europe, the Canary Islands, the Middle East, and Africa. The airline operates a large number of package tour flights between resort areas of Spain and cities of Central and Northern Europe, as well as the United Kingdom. A fleet of DC-9 and F-27 aircraft is used. *(Udo and Birgit Schaefer Collection)*

Above: **Balair** (BB)
Balair, Switzerland's largest charter airline, was formed in 1953, with flight operations commencing four years later. The carrier operates worldwide passenger and freight flights with a fleet of jet aircraft. Balair ownership is made up by Swissair (57 per cent), other private interests (31 per cent), and the Swiss government (12 per cent). A majority share of Balair flights involves the transportation of tour groups over intra-European and intercontinental routes to and from Basle, Geneva and Zürich. Regular charters serve North Africa, Kenya, Togo, Gambia, Sri Lanka, the Maldives and USA with a fleet of DC-9, DC-10, MD-80 and A310 aircraft. One of the company's two MD-80s is seen here. *(Udo and Birgit Schaefer Collection)*

Opposite: **British Midland Airways** (BD)
British Midland Airways was founded in 1938 as Derby Aviation, initially operating as an air training school. In 1953 the airline commenced scheduled services to Amsterdam, Brussels and Paris (CDG) from its base at East Midland Airport. Domestic flights operated from many regional airports throughout the UK. Since the end of 1982 the airline has been engaged in direct competition with British Airways on major domestic routes out of London (Heathrow) Airport. Multiple non-stop DC-9 flights are operated to Belfast, Edinburgh and Glasgow. In 1986 new daily DC-9 services were introduced on the highly competitive route between London (Heathrow) and Amsterdam. Additional services are also operated to Leeds/Bradford, Birmingham, East Midlands and Teesside. BMA has a 75 per cent shareholding in Manx Airlines and also Loganair, the Scottish regional airline. The company operates a fleet of DC-9, F-27 and Shorts 360 aircraft. *(P Hornfeck)*

BM
British Midland
G-BMAA
BM
HALT

CAAC-General Administration of Civil Aviation of China (CA)
CAAC is the government-controlled national flag carrier of the People's Republic of China. The airline was established in 1962 as the General Administration of Civil Aviation of China. In addition to its airline operations, the carrier controls all civilian air transport activities in China (including passenger, cargo, agricultural, airport training and other specialised services). CAAC maintains a scheduled domestic route network and serves international destinations in Asia, the Middle East, Africa, Europe, the USA and Australia. The carrier operates a fleet of Boeing 707, 737, 747, MD-80, Il-62, Trident and BAe 146 aircraft. Besides these types, various turboprop aircraft are also utilised. *(McDonnell Douglas)*

Continental Airlines (CO)
Continental Airlines, a subsidiary of the Texas Air Corporation, was formed in 1934, when Varney Space Lines commenced Lockheed Vega servvices between El Paso, Albuquerque and Pueblo. In 1936, Varney purchased a Denver-Pueblo route from Wyoming Air Service, changed its base to Denver, and renamed itself as Continental Airlines. Today the airline operates scheduled flights to the South Pacific, Central Pacific, East Asia, Mexico, Canada, and the UK. The carrier also operates regional Pacific flights on behalf of Air Micronesia. Continental has a fleet of A300B4, Boeing 727, 737, DC-9, DC-10 and MD-80 aircraft. A DC-9-32, N521TX, is seen here taxiing after another flight. *(Udo and Birgit Schaefer Collection)*

Cruzeiro Brazilian Airlines (SC)
Although no longer operating the MD-80, Cruzeiro used this type on many domestic services. The company was created in 1927 as Kondor Syndikat. Initial services linked Rio de Janeiro and Porto Alegre with other southern Brazilian points. Today, Cruzeiro, Brazil's third largest airline, operates a scheduled jet service route network to over 26 points within its home country, as well as serving destinations in Argentina, Barbados, Bolivia, French Guiana, Peru, Surinam and Trinidad and Tobago. A fleet of Boeing 727, 737 and A300 aircraft is currently utilised. *(McDonnell Douglas)*

Delta Airlines (DL)

Delta Airlines was established as a crop dusting company in 1924, with a base at Macon in Georgia. The carrier took the title of Delta Air Services in November 1928, with scheduled passenger operations beginning some seven months later. The inaugural flight was made over a Dallas-Shreveport-Monroe-Jackson route. Today Delta ranks high among the world's airlines in terms of passenger carriage. The company schedules well over 1000 daily departures to over 90 points in the USA, Canada, Bermuda, the Bahamas, Puerto Rico, and Europe. Over 200 aircraft are utilised. With a base at Atlanta, Delta is widely regarded as one of the world's most efficiently-run airlines. A fleet of L-1011, Boeing 727, 737, 757, 767, DC-8 and DC-9 aircraft is used. The company has a number of MD-88s on order to replace its ageing fleet of DC-9s. *(Udo and Birgit Schaefer Collection)*

Below: **Eastern Airlines** (EA)
Eastern Airlines was formed in September 1927 as Pitcairn Aviation, an airmail carrier. Pitcairn began mail runs over a New York-Atlanta run a year later. In 1929 the company became Eastern Air Transport. Today, Eastern Airlines is one of the world's largest air carriers. The company maintains an extensive scheduled network that connects over 80 points across the USA, as well as over 37 foreign destinations in Canada, the Bahamas, Bermuda, the Caribbean, Mexico, Central and South America. Eastern operates a huge fleet of L-1011, Boeing 727, 757, A300 and DC-9 aircraft. A DC-9-31, N8983E, is depicted here.
(Udo and Birgit Schaefer Collection)

Opposite: **Emerald Air** (OD)
Emerald Air suspended all operations on 21 August 1984, and filed for financial reorganisation under Chapter 11 of the Federal bankruptcy laws. Privately-owned, the carrier was formed in 1978. Scheduled contract flights for Purolator Courier began two years later, with scheduled passenger services beginning in 1981, utilising turboprop aircraft. A route structure covering the Midwest and Texas was maintained. Throughout mid-1984 the carrier had linked up with Pan Am to provide an Emerald Air/Pan Am Express. Services were made to and from a hub at Houston Intercontinental Airport. At the time of the demise, a fleet of DC-9, FH-227 and Gulfstream 1 aircraft was utilised.
(R H Vandervord)

Emerald Air
N3864L

FINNAIR
MD-83
MD-83

Opposite: **Finnair** (AY)

Finnair, the government-controlled national airline of Finland, was formed in 1923, when Bruno L Lucander established Aero o/y. An inaugural flight was made between Helsinki and Tallinn in Estonia on 20 March 1924, using a four-seater Junkers aircraft. Following the Second World War, the carrier became Aero o/y Finnish Air Lines, and subsequently Finnair. The airline has a 60 per cent interest in Finnaviation, and a 35 per cent holding in Karair (Finland), as well as operating various Finnish travel agencies, tour firms and hotel services. The carrier maintains scheduled services over an extensive network in Finland, as well as to international destinations throughout Europe, North America, North Africa and the Far East. Finnair operates a fleet of ATR 42, DC-9, DC-10, MD-80 and MD-83 aircraft. An example of the latter type can be seen here. *(McDonnell Douglas)*

Below: **BWIA International** (BW)

The McDonnell Douglas MD-82 pictured here was leased from Frontier Airlines between June 1985 and March 1986. BWIA now operates the MD-83 series in its place, this type being leased from Guinness Peat Aviation. Although the company operates to international points, BWIA covers a large domestic network that links the many islands in the West Indies. With a base on the island of Trinidad, the airline employs over 1000 people worldwide. *(N Mills)*

Above: **Hawaiian Air** (HA)
Hawaiian Air was established as Inter-Island Airways in 1929, and commenced scheduled flights between Honolulu and Hilo in November of that year using Sikorsky S-38 amphibious aircraft. In 1941 the company's title was changed to Hawaiian Airlines, and DC-3 equipment was introduced. Today the airline schedules over 130 daily departures from seven airports on six Hawaiian islands, as well as providing international services to American Samoa and Tonga. Hawaiian Air undertakes tour group charter and Hawaiian scenic tours. A fleet of L-1011 Tristar, DC-9, MD-80 and de Havilland Canada Dash 7 aircraft is operated. A DC-9-51 is seen here. *(Udo and Birgit Schaefer Collection)*

Opposite: **Iberia** (IB)
Iberia, Spain's government-controlled national airline, was formed in June 1927 and began flights in December of that year over a Madrid-Barcelona route. The company operates scheduled passenger flights to over 20 destinations in mainland Spain, the Balearic Islands, Melilla and the Canary Islands. Iberia also serves over 60 points in Europe, Africa, the Middle East, North, Central and South America, as well as the Caribbean. The carrier's investments include a minority shareholding in Aviaco. Iberia operates a fleet of A300B4, Boeing 727, 747, DC-9 and DC-10 aircraft. A DC-9-32 is seen here at London (Heathrow) Airport. *(J Page)*

IBERIA
EC-BPH

Intercontinental de Aviacion (RS)

Intercontinental de Aviacion was established in October 1965 as Aeropesca Colombia. The privately-owned airline changed to its present name in December 1982, and subsequently supplemented Viscount aircraft with DC-9 equipment. Today, the carrier provides scheduled passenger jet services to 12 points throughout Colombia. Intercontinental operates a fleet of Curtis C-46 freighters, Viscount and DC-9-15 aircraft.
(Udo and Birgit Schaefer Collection)

JAT Yugoslav Airlines (JU)
JAT is the state-run flag carrier of Yugoslavia, and is one of Eastern Europe's most prominent passenger carriers. Equipped with an all US-built aircraft fleet of Boeing 707, 727, 737-300, DC-9 and DC-10 types, the airline maintains comprehensive scheduled services in Yugoslavia. In addition it covers international routes throughout Europe and to the Middle East, North Africa, Southern Asia, Australia and North America. JAT was formed in April 1947 to take over from its predecessor, known as Aeroput. In addition to scheduled operations, the Yugoslav airline undertakes extensive charter services, some through a charter subsidiary known as Air Yugoslavia. Air taxi flights are also handled by a JAT Aviotaxi division. A DC-9-32 is seen here at Copenhagen Airport.
(Udo and Birgit Schaefer Collection)

Jet America (SI)
Jet America was formed in September 1980, with an inaugural flight being made between Long Beach and Chicago on 16 November of the following year. Until its demise in 1986, the carrier was flying scheduled passenger services between California, the Midwest and Texas. Publicly-held Jet America also undertook group charters and contract flying. The airline operated a fleet of modern MD-80 aircraft from its base at Long Beach, California. N783JA is seen here at Las Vegas during June 1986. *(R H Vandervord)*

KLM — Royal Dutch Airlines (KL)
KLM, the national flag carrier of the Netherlands, is the world's oldest airline. The company was formed in 1919, with an inaugural service between Schiphol and London on 17 May 1920. The carrier maintains one of the world's most extensive international systems, connecting well over 100 cities in Europe, Africa, the Middle East, southern and eastern Asia, Australia, North, South and Central America, as well as the Caribbean. KLM is owned by private investors and the Dutch government, and itself owns several subsidiaries that include KLM Aerocarto, KLM Helicopters and NLM City Hopper. Additional interests are held in Martinair Holland, XP Parcel Express Systems, Schreiner Aviation Group, and in hotel, travel and bus companies. The airline operates a fleet of Boeing 737, 747, DC-9, DC-10 and A310 aircraft. KLM's DC-9 fleet is due for replacement by the 'new generation' Boeing 737-300 aircraft.
(Udo and Birgit Schaefer Collection)

Korean Air Lines (KE)
KAL was formed in 1962 by the South Korean government to take over the operations of Korean National Airlines. In 1969 ownership was transferred to the private Hanjin Group, a business organisation that has holdings in many varied financial undertakings. In addition to its airline operations, Korean Air is involved in fighter aircraft and helicopter production activities, primarily for the South Korean Air Force. The airline operates scheduled services to 36 points in South and Eastern Asia, North America, Europe and North Africa. Korean Air maintains a fleet of Boeing 707, 727, 747, DC-10, MD-80, A300, F-27, F-28 and CASA 212 Aviocar aircraft. The spectacular new corporate livery is well displayed on this newly delivered MD-80. *(McDonnell Douglas)*

Martinair Holland (MP)
Martinair Holland is one of Europe's leading charter airlines. From its base at Amsterdam Schiphol Airport, the carrier operates worldwide passenger and freight charter and contract services. The company was established in 1958 as Martin's Air Charter, and commenced flight operations with a DC-3 aircraft. Ownership control is held by the Royal Nedlloyd Group (49 per cent), KLM Royal Dutch Airlines (25 per cent), as well as several banking and investment companies. The Martinair Holland group also includes food and catering services, advertising, flight training, and the sales and maintenance of business and sport aircraft. A fleet of MD-80, DC-10 and A310 aircraft is utilised. *(Udo and Birgit Schaefer Collection)*

Midway Airlines (ML)

Midway Airlines was formed in 1976 and commenced flights in November 1979 between Chicago Midway, Cleveland, Detroit and Kansas City. The company had the distinction of being the first major new US carrier to begin operations following the passage in 1978 of the Airline Deregulation Act. Through mid-1983 Midway was developed as a low-fare carrier. In June of the same year, in an effort to stimulate traffic, the carrier introduced the Midway Metrolink service, designed for frequent business travellers. This was the first offered over the Chicago-New York route, and proved to be such a great success that it was expanded system-wide. The airline operates a route structure that covers the Midwest, the East and Texas. A fleet of Boeing 737, DC-9 and MD-80 aircraft is utilised. A DC-9-31 is seen sporting the Midway livery.
(Udo and Birgit Schaefer Collection)

Midwest Express (YX)
Midwest Express gained approval from the CAB in November 1983, and on 11 June 1984 began operating DC-9-14 jet flights. The company operates an all 'business class' system of flights in the Midwest, the Northeast, and to Texas. The carrier is a subsidiary of K-C Aviation (itself owned by Kimberley-Clark, a major paper products manufacturer). In addition to its public scheduled services, the carrier also undertakes company shuttle flights for Kimberley-Clark. The airline serves Appleton, Boston, Dallas, Milwaukee and Newark.
(Udo and Birgit Schaefer Collection)

Muse Air
N932MC
Muse Air

Opposite: **Muse Air Corporation** (MC)
Muse Air was formed in January 1980 by Lamar Muse, former President and Chief Executive Officer of Southwest Airlines. An inaugural service flew between Dallas and Houston on 15 July 1981. The company expanded its operations very quickly, and scheduled low-cost passenger services across the Central and Western Sun Belt, linking cities in Texas, Oklahoma, Lousiana, Nevada and California. An unusual touch introduced by the airline was that all services were 'No-smoking'. In January 1986 the company changed its name to Transtar. At that time a mixed fleet of DC-9 and MD-80 aircraft was operated. *(S Ahn)*

Above: **Northeastern International Airlines** (QS)
Northeastern operated scheduled low-fare passenger flights along the eastern seaboard between the Northwest and Florida as well as between the Midwest and Florida. The company was formed in March 1980, and gained authority for charter flights in December of that year. Services began over an Islip-Ft Lauderdale line, using DC-8 equipment. Unfortunately, in January 1985, due to escalating difficulties, the airline filed for bankruptcy. At the time, a fleet of A300, Boeing 727, DC-8 and DC-9 aircraft was utilised. MD-80 types were to have been leased from Alisarda during the winter months, although this never actually materialised. *(N Mills)*

Northwest Airlines (NW)
Northwest Orient was formed in August 1926 as Northwest Airways, and initially operated mail flights between Minneapolis and Chicago. The current name was adopted in 1934. The airline maintains scheduled flights to nearly 100 destinations in the USA, as well as serving Canada, the Pacific, Eastern Asia and Europe. Late in 1984 the company set up 'Northwest Orient Airlink'. This is an agreement with various independent and regional air carriers to feed passengers from the company's main routes. In 1986 Republic Airlines was purchased by Northwest and was merged into the company. October of the same year saw Northwest place an order for as many as 100 Airbus A320 aircraft. A fleet of DC-9, DC-10, Boeing 727, 747, 757, CV-580 and MD-80 types is utilised. *(N Mills)*

Ozark Airlines (OZ)
Ozark Airlines provides scheduled passenger jet services to over 50 destinations within the USA. The carrier's main flight hub is in St Louis, Missouri, with extensions to the eastern seaboard, the Southeast, Florida, Texas, the Rocky Mountains, Nevada and California. The airline commenced operations in 1950 over a St Louis-Springfield-Decatur-Champaign-Chicago route, initially utilising a fleet of DC-3 aircraft. Today, Ozark operates a fleet of over 40 DC-9 and MD-80 types. Besides the scheduled services, the airline operates regular group charters within the USA, Canada, Mexico and the Bahamas. A Douglas DC-9-40 is pictured here while taxiing.
(Udo and Birgit Schaefer Collection)

Pacific Southwest Airlines (PS)
Pacific Southwest Airlines was established in 1949, with an inaugural flight on 6 May of that year using a single Douglas DC-3 over a San Diego-Burbank route. Until 1978 PSA operated entirely within the state of California, but subsequent deregulations led to a substantial increase in interstate expansion. In 1980 international services to Mexico were inaugurated, but these were discontinued two years later. The airline schedules well over 300 daily passenger flights along a regional route system that links more than 20 points in California, Arizona, Nevada, New Mexico, Oregon and Washington. PSA operates a fleet of MD-80, DC-9 and BAe 146 aircraft. An example of the MD-80 can be seen here in flight in full PSA livery. *(McDonnell Douglas)*

Purolator Courier

Purolator Courier is a major United States small package express and freight forwarding company which operates nationwide pick-up and delivery services. The carrier utilises full or partial space on well over 100 different aircraft, many of which are contracted from various independent companies, such as the large express freight companies. Purolator undertakes overnight flights throughout the USA. A central package sorting hub is located at Indianapolis. A number of DC-9, Boeing 727 and prop aircraft make up part of a very large fleet. N562PC, DC-9-15, is seen here at Minneapolis/St Paul in June 1986. *(R H Vandervord)*

Republic
N8907E
Republic

Opposite: **Republic Airlines** (RC)
Republic Airlines was formed on 1 July 1979, following the merger of North Central Airlines and Southern Airways. In 1986 the carrier was purchased by Northwest Airlines and all services have now been integrated with those of the parent company. Prior to this, Republic was one of the largest companies in the USA, serving well over 100 destinations across the USA, Canada, Mexico and the Cayman Islands. From its base at Minneapolis/St Paul, Republic operated a fleet of DC-9, Boeing 727, Boeing 757 and CV-580 aircraft. A Douglas DC-9-10, N8907E, is seen here at the company's home base.
(Udo and Birgit Schaefer Collection)

Below: **Sunworld International Airways** (JK)
Sunworld International, a privately-owned carrier, was founded on 4 June 1981 as Jetwest International Airways. In early 1983 the company adopted its present title and obtained operating authority from the Civil Aeronautics Board. The carrier's inaugural flights were made on 27 May of that year between Las Vegas, Ontario and San Jose. Today the company operates scheduled passenger flights to cities in the Southwest and California, from a hub at Las Vegas. A fleet of Boeing 737-300 and DC-9-14 is operated. N1302T is seen here at the company's operational base at Las Vegas.
(R H Vandervord)

swissair
HB-IFU

Opposite: **Swissair** (SR)
Swissair was founded in March 1931 via the merger of Ad Astro Aero and Basle Air Transport. The company was in 1947 designated as national flag carrier. From then, rapid expansion followed in scheduled operations. Swissair is mainly privately-owned, with public institutions holding the remaining shares. The airline has itself an interest in Balair, in addition to holdings in CTA, two Swiss charter companies. Today the carrier maintains scheduled jet services over a vast worldwide network. The company owns a fleet of Boeing 747, DC-9, DC-10, MD-80 and Airbus A310-200 and -30 types. The entire DC-9 fleet, of both -32 and -51 marks, will be phased out to make way for the all-new Fokker 100.
(Udo and Birgit Schaefer Collection)

Above: **Texas Air Corporation**
New York Air is part of the Texas Air Corporation, a huge conglomerate which owns People Express and Continental, among others. Based at Houston, the company has an 85 per cent interest in Continental Airlines and wholly owns the New York-based airlines. The company has a fleet of DC-9, MD-80 and Boeing 737 aircraft.
(Udo and Birgit Schaefer Collection)

Below: **Texas International Airlines** (TI)
Texas International, until 1969, was known as Trans-Texas Airways. The company was established in 1940 as Aviation Enterprises, and began scheduled local services in October 1947. TXIA operated scheduled passenger and cargo services to points in nine states, stretching from Los Angeles in the west to Washington DC in the east. International services were operated to Cozumel, Cancum, Merida, Mexico City, Guadalajara and Monterrey in Mexico. On 28 October 1982, the airline was merged with Continental Airlines. *(D Penny)*

Opposite: **THY Turk Hava Yollari — Turkish Airlines** (TK)
THY, the government-controlled flag carrier of Turkey, was formed in 1953 as Devlet Hava Yollari, with the present name being adopted in 1956. The government has a majority shareholding in the carrier, which provides scheduled services within Turkey, as well as international routes that cover Europe, North Africa, the Middle and Far East, as well as Pakistan. THY has subsidiary companies that include Cyprus Turkish Airlines. The company has its major international centres at Istanbul's Yesilkoy Airport and Ankara's Esenboga Airport. THY operates a fleet of Boeing 707, 727, DC-9 and A310 aircraft. A DC-9-15, no longer operated, is seen here in flight. *(McDonnell Douglas)*

TC-JAA

TDA
東亜国内航空 TDA
MD-81

Above: **Touraine Air Transport** (IJ)
TAT, the largest of the French regional airlines, was established in 1968. Scheduled services commenced two years later. The company maintains regular scheduled passenger services throughout mainland France and to Corsica. Additional operations include numerous flights on behalf of Air France. Although DC-9s are no longer operated, they were leased from Finnair, and were used to destinations that included London (Heathrow). The current fleet consists of F-27, F-28, FH-227, DHC-6 Twin Otter and Beech 99 aircraft.
(Udo and Birgit Schaefer Collection)

Opposite: **Toa Domestic Airlines — TDA** (JD)
Toa Domestic, Japan's second largest internal carrier, schedules more than 300 daily departures from over 38 destinations along an extensive route system that covers the major islands of Honshu, Hokkaido, Shikoku and Kyushu, as well as Nansei (Southwest) Islands. The airline is owned by the Tokyo Express Electric Railway Company, Toa Kosan Ltd, JAL, Kinki Nippon Railway Company, and other interests. Toa Domestic was formed on 15 May 1971 through a merger of Japan Domestic Airlines (JDA) and Toa Airways (TAW). The airline has itself a 60 per cent interest in Japan Air Commuter, a carrier serving routes in the Nansei Islands. Toa operates a mixed fleet of aircraft ranging from A300B2, MD-81 and DC-9s, to ageing NAMC YS-11 and numerous helicopters.
(McDonnell Douglas)

Transtar Airlines

Transtar Airlines, until January 1986, was known as Muse Air. Operations commenced in July 1981 using MD-80 aircraft. With an initial flight between Dallas and Houston, services were increased to include Oklahoma City, Los Angeles, San Diego, New Orleans, Orlando and Tampa. The current company name was officially adopted in February 1986, when Southwest Airlines purchased the company. A fleet of DC-9 and MD-80 aircraft is operated, with MD-80, N935MC, seen here at Las Vegas. *(R H Vandervord)*

Trans World Airlines (TW)
TWA, one of the world's most prominent air carriers, maintains an exhaustive route network of jet services that link over 60 points across the USA, as well as over 20 foreign destinations in Europe, the Middle East and India. The history of TWA can be traced back to 1925, with the formation of Western Air Express. Following a merger in 1930 with Transcontinental Air Transport, the company became Transcontinental Western Airlines. The current name was adopted in 1950. In 1984 as a result of a separation from the Trans World Corporation, TWA became an independent publicly-held company. The airline operates a fleet of Boeing 727, 747, 767, L-1011 and MD-80 aircraft.
(Udo and Birgit Schaefer Collection)

USAir (AL)
USAir maintains over 1000 daily flights across the USA. In 1968 Lake Central Airlines was absorbed by the then Allegheny Airlines, which extended routes into the eastern Midwest. Mohawk Airlines was merged in April 1972, which extended and consolidated routes throughout the Northeast, thus making Allegheny the USA's premier regional carrier. On 28 October 1979 USAir became the airline's new name. N980VJ, a DC-9-32, is seen here at Minneapolis/St Paul, sporting the carrier's current livery. *(R H Vandervord)*

VIASA (VA)
VIASA, the Venezuelan nationl flag carrier, provides scheduled services along an intercontinental route network that covers Central, North and South America, the Caribbean, and Europe. The company was established in November 1960 to take over most international services of Aeropostal (LAV) and Avensa, airlines which had 55 per cent and 45 per cent shareholdings in this new company respectively. The carrier's official inaugural flight was between Caracas and New York, utilising a newly-delivered Convair 880 aircraft. Currently VIASA operates a fleet of DC-10 aircraft.
(Udo and Birgit Schaefer Collection)

Above: **Douglas DC-9**
In 1962 Douglas announced its intentions to build a twin-engined commercial jet airliner. On 25 February 1965, the DC-9-10 first took to the skies. N9DC and its Series 10 companions were designed to carry up to 90 people and transport them a distance of 2110 km (1311 n miles). Since that first example flew, the aircraft fuselage has been stretched many times, and the range increased. The latest in the Douglas family is the MD-83, which can fly distances of over 4635 km (2501 n miles). *(McDonnell Douglas)*

Opposite: **AeroMéxico** (AM)
In the old livery of AeroMéxico, this DC-9-15 'short body' is seen here taxiing into a Mexican airport. The airline was established in September 1934 as Aeronaves de Mexico SA. In the 1940s and 1950s various other Mexican carriers were absorbed by Aeronaves, including Aeronavias Michacan-Guerrers, Taxi Aéreo do Oaxaca, Lineas Aéreas Jesus Sarabia, LAMSA and Aerovias Reforma. In 1957 an important international expansion took place when Aeronavias introduced Bristol Britannias over its Mexico City-New York route. It was not until the late 1960s that the carrier introduced DC-9 equipment. *(D Penny)*

AEROMEXICO
XA-DEV

Hawaiian Air (HA)
This Hawaiian Airlines DC-9-15 is seen here during March 1966, while sporting the company's initial colour scheme. Although the company no longer operates this version, Hawaiian has always been actively involved with McDonnell Douglas and has operated the -32, -51, and MD-80 aircraft types. Douglas DC-8s have also been operated. Today, MD-80s operate the Hawaiian Island 'air-link', which connects the seven islands (Hawaii, Kahoolawe, Kauai, Lanai, Maui, Molokai and Oahu). *(McDonnell Douglas)*

Iberia (IB)

Seen here sporting Iberia's original livery, DC-9-32, EC-BPG, is pictured arriving at London (Heathrow) Airport. The Spanish national flag carrier was formed on 28 June 1927. The carrier operates a worldwide international network and utilises a fleet of aircraft dominated by American manufacturers. Iberia operates an air shuttle 'Puerto Aéreo' operation providing high frequency daily non-stop flights between Madrid and Barcelona.
(Udo and Birgit Schaefer Collection)

Below: **Midway Airlines** (ML)
Using a DC-9 in the livery of Midway Metrolink, Midway Airlines introduced a new look in June 1983. Designed for frequent travellers, the carrier offered a brand new service on a Chicago-New York route. With newspapers, beverages and additional baggage capacity, the company was able to provide that extra touch for the business market. This triggered a great increase in traffic and so popular was the service that the idea was extended to the company's other routes.
(Udo and Birgit Schaefer Collection)

Opposite: **USAir** (AL)
USAir, a subsidiary of the USAir Group, was formed as All-American Aviation in March 1937, and began airmail flights in West Virginia and Pennsylvania two years later. In 1949, having changed its name to All-American Airlines, the company started scheduled passenger flights utilising DC-3 equipment. In 1953, another name change to Allegheny Airlines was made. The company maintains a comprehensive scheduled passenger network in the Northeast, eastern Midwest, Florida, and many other areas including Canada. The airline operates over 1000 daily flights. USAir has a base at Washington National Airport, and has a main flight base and maintenance centre at Greater Pittsburgh International Airport. The airline operates a fleet of Boeing 727, 737, DC-9 and BAe One-Eleven aircraft. A DC-9-30 is seen here sporting the old Allegheny colours. *(McDonnell Douglas)*

ALLEGHENY
N920VJ

Austrian Airlines (OS)

Austrian Airlines, the flag carrier of Austria, is almost entirely state-owned. The airline was formed in September 1957 by combining two airlines, Air Austria and Austrian Airways, before either had commenced operations. An inaugural flight was made on 31 March 1958, using a leased Viscount aircraft over a Vienna-London (Heathrow) line. It was not until 1963 that the airline introduced jet equipment in the form of a Caravelle. Austrian Airlines has a base at Vienna and flies to over 42 destinations in Europe, North Africa and the Middle East. The carrier operates an all-McDonnell Douglas fleet, consisting of DC-9-32, DC-9-51 and MD-80 aircraft. MD-87 types will be introduced in late 1987, while two A310-300 aircraft will follow in 1988 and 1989. A DC-9-32 is seen here, while taxiing.
(Udo and Birgit Schaefer Collection)